Earth Resonance

Earth Resonance
Poems for a Viable Future

⌘

Sam Love

A Publication of The Poetry Box®

Poems ©2022 Sam Love
All rights reserved.

Editing & Book Design by Shawn Aveningo Sanders
Cover Photograph by Sarah Granger
Cover Design by Shawn Aveningo Sanders

No part of this book may be reproduced in any manner whatsoever without permission from the author, except in the case of brief quotations embodied in critical essays, reviews and articles.

ISBN: 978-1-948461-86-3
Library of Congress Control Number: 2021921431
Printed in the United States of America.
Wholesale Distribution via Ingram.

Published by The Poetry Box®, March 2022
Portland, Oregon
ThePoetryBox.com

Resonance

A small vibration at the right frequency can create a larger vibration. For example, if air is forced inside a conch shell and begins to vibrate through the spiral chamber, a resonance is created, and the result is a big vibration, a loud sound. So, let the voice of a poet resonate to expand our awareness of our impact on the planet and pave the way to a viable future.

Contents

Planetary Unconsciousness

Golden Spiral	13
Our Legacy	14
Turtle Earth	15
The Web	16
If We Could Hear the Earth	17
Monarch's Lament	18
Primal Memory	20
Jacuzzi Guilt	21
Crazy Water	23
Blurred Vision	25
Urban Tumbleweeds	27
Spermageddon	28
Star Man	29
Another Planet	30
Happy Birthday Nature	31
Drowning in the Internet	32
Mother Earth Meditation	33
Rebels of the Universe	34
Each Day	35

Culinary Consciousness

Viva La Tomato	39
Karmic Revenge	40

Blueberry Mourning	41
The Perfect Holiday Meal	43
Cooking Green	44
Winter Smorgasbord	45
The Downstream Loop	46
Sayonara Humans	47

Nature's Canaries

The Last Laugh	51
Deaf	52
Elegant Travelers	53
Translucent Canaries	55
Nature's Bacchanal	57
Lightning Bug Apology	58
Leaf Meditation	59
Hummingbird Laughter	60
Mourning Dove Spring	61
Lunacy	62
Ghost Stumps	63
Long Leaf Cathedral	65
Chasing Venus	66
Riding It Out	68
Mother Nature Is a Bitch	70
Recovering Hope	72
Unleashing the Cosmic Jokers	74

Deconstructing the Future

Awakening	79

A Tiny Acorn	80
Forest Bathing	81
Homo Sapiens' Abridged History	82
Highway Dreams	83
A Monument to Another Time	85
Crossing the Void	86
Visiting Khaytyn	87
Redefining Habitat	88
Ballet Mechanistic	90
Big Blue Monster	91
How Many Falls?	93
The Rebellion	95
Grass Meets Its Match	96
Apologies to a Tree	97
Grey Dawn	98
Evolution's U-Turn	99
Monkey Mind	100
Restoring Consciousness	101
Sitting Still	102
Reshaping the Future	103
Story Power	104

Earth's Epilogue

Rust	107
Rusted Dreams	108
One Word: Plastic	109
City in Grey	110
Nearly Natural	111

Forever Green	112
The Ecology Symbol	114
Unleash the Dove	116
A Perfect Legacy	117
Acknowledgments	119
About the Author	121
About The Poetry Box®	123

Planetary Unconsciousness

Golden Spiral

Blowing the conch shell
can herald a call to prayer,
warn of danger,
or celebrate a victory.

With lips clinched the breath
enters a small hole
in the end of the shell
and bounces off the spiral cavity
to expand in volume.

The conch mimics nature's designs
of spiral galaxies, spiral bacteria,
packed atomic particles,
and the contours of sand dunes.

This sympathetic vibration
resonates inside the conch
until a loud ohm-like sound
exits the large opening.

Listen to contemporary poets
trumpet warnings of:
global warming,
stronger hurricanes
increased forest fires,
and beaches so hot
mollusks cook in their shell.

May this voicing of our survival instinct
resonate like the expanding volume
in the conch and awaken the masses.

Our Legacy

As obsolescence
rules the market
permanence becomes
a museum exhibit.

Native American potters
must be smiling,
knowing their earthen art
will outlast our digital legacy.

Still, our legacy is secure
in mounds of disposables
no one can repair
and nature cannot degrade.

A legacy archived
in plastic bits and
chemicals polluting
the Earth's waters.

Turtle Earth

In the Lenape creation story—Nanapush asks,
"Who will let me put cedar branches on top of you
so that all the animals can live on you?"
And the turtle says, "You can put them on me
and I'll float on the water."

In a vision the Native American holy man
sees the animals bringing earth
from under the water to make land
on the back of the turtle
to create a verdant Eden
where plants and animals flourish.

In another dream the Indian shaman
sleeps a long sleep and
sees a barren turtle
with writhing serpents
thrashing rattlers through portals
in its armor-plated shell.

This hollow eerie sound
resonates with a dry rattle
of primordial notes memorializing
the emerging death of nature.

The Web

No one is alone,
we are all part
of life's web.

With each breath
we inhale remnants
of the universe and exhale
nourishment for plants.

Breaking one thread
of the food web—
like poisoning the bees
harms pollinating plants
and disrupts our food supply.

Still we continue to throw
currencies into the wind
to purchase reality.

Instead we should realize
we are only one strand
in the delicate web.

If We Could Hear the Earth

~a children's poem

When we punch holes
to drill for oil and gas
the planet will say
ouch!

As we clear cut
oxygen producing forests
the sky will
cough!

As monster shovels dig
giant mines
the mountains will
cry!

As we release CO_2
to warm the planet
thunderstorms will
rumble!

As we disrupt the ocean
with disposable plastic
the fish will
SCREAM!

If Mother Earth could speak
she would tell us
to change our path
or prepare to be homeless.

Monarch's Lament

-a tribute to Homero Gómez González and Raul Hernandez Romero, Mexican Monarch Guardians murdered in 2020

As this summer's fourth generation
our chrysalis cocoons now hatch
to prepare for our southern migration
to our ancestral forest.

As the days shorten
our antennae sense the changes
and we rise skyward
in search of a southern current
to boost our flight.

With only a few rest stops we
follow rivers and mountains
thousands of miles south
to a Mexican mountain forest.

As our cloudlike masses
pass over natives
they worship us as
spirits of returning warriors.

Our genetic imprint
guides us to the trees
where our ancestors hatched.

Arriving by November's Day of the Dead
we find many of our Oyamel fir trees
have been cut leaving us
a barren wasteland.

And our brave guardians led by
Homero Gomez and Raul Hernandez
gave their lives fighting
the illegal forest loggers.

After hibernating, rising temperatures
signal it's time
to fly north to breed
and create another generation.

Before taking wing
some of us join mourners
inside a service honoring
Homero Gomez González,
one of our guardians.

After we flutter inside the chapel
we start our journey north
and wonder who will guard
our forest guardians?

Note: As of April 2021, no one has been arrested for the murders of Homero Gomez and Raul Hernandez. Global Witness reports an average of four defenders of habitat and water have been killed every week since the creation of the Paris Climate Agreement in December 2015.

Primal Memory

Near the mouth of Swift Creek
a gnarled head bobs up
flips its tail and
swims away.

As I power my boat closer
ancient eyes briefly gaze at me
with the genetic memory

of a creature whose ancestors swam
with dinosaurs and escaped
extinction for 65 million years.

Now it faces a new threat
as decades of wildlife protection
come to an end

and alligator hunting
becomes legal again
in North Carolina's rivers.

The next time I see the ancient gator
the skin of its belly could
be covering $2,000 custom boots.

Jacuzzi Guilt

As I unleash six water jets
in my spotless whirlpool tub,
the warm water quiets a symphony
of snaps, crackles and pops
in my aging joints.

My body relaxes but
my mind churns
as furiously as the bubbles
in the tub's sixty gallons.

I soak in as much
water as a rural African
villager uses in ten days.

As my body unwinds
my marauding mind contemplates
my contribution to
global warming.

Does the pump's electricity
release greenhouse gases
from strip-mined coal?

Or was the power generated
by natural gas from
a leaking fracked well?

Did the tub's water flow
through an upstream flooded
animal waste lagoon before
chlorine sanitized the liquid?

[. . .]

What is the carbon footprint
of my favorite Dead Sea salts
that fly half-way round the planet
to dissolve in my bath?

As clean water becomes scarce
how can the earth's population
afford such affluence?

But as the salted water
relaxes my critical mind
I dismiss any thoughts
of how my luxurious lifestyle
cripples the planet's future
and I just enjoy the bath.

Crazy Water

A Tale of Two Raindrops

Nothing says affluence more
than a square bottle of pure water
from a virgin South Pacific ecosystem
where raindrops fall on a pristine rainforest.

Empty American plastic
bottles travel thousands of miles
to Fiji for filling from
the mineral-laden volcanic aquifer.

Machines twist on Taiwan plastic caps
and apply labels
printed who knows where.
Once filled and ready,
the square bottles are snugged
into an imported cardboard box.

Next, the bottles load into shipping containers,
so the water can move 6,000 miles
across the Pacific to ports in California
or through the Panama Canal to the East Coast
where cranes like giant erector sets
unload the containers of water
for shipping to your store.

[. . .]

Savoring each smooth sip evokes
the advertised illusion of health,
but what's wonderful for your body
is not wonderful for the planet
as one million Fiji water bottles
per day move around the world
on belching container ships.

To really enjoy the Pacific's finest water
we must ignore how its carbon footprint
contributes to climate change's
record hurricanes and floods

and realize many communities
can never afford
to grip a crystal-clear
square plastic bottle.

Blurred Vision

We seem oblivious
that each of our little acts
blurs our vision of
our children's future.

Each little act
like flushing contact lenses
that confuse fish who think
they are delectable fish eggs.

Each little act
like cutting weeds
with a weed whacker
that breaks off little plastic strings
that disappear down the drain.

Each little act
like walking past
trashed plastic bags poised
to wash into the river.

Each little act
like taking one
of the 90,000 flights per day
to visit grandchildren
and ignoring the plane's
contribution to global warming.

[. . .]

To clearly envision
a future without
melting glaciers
stronger hurricanes
and boiling summers,

We need to contemplate
how each little act
can be a gift to our grandchildren.

Urban Tumbleweeds

Like a Western tumbleweed
a synthetic urban tumbleweed
bounces across a highway,
revealing orange splashes of sunlight
shimmering through the plastic bag.

Gyrating in wind-driven swirls,
the bag dances across the road,
until it catches on the curb, deflates
and waits for the next rainstorm
to wash it down the storm drain.

Where it floats out of sight
until surfacing in the river
to join other plastic detritus
destined to become the legacy
of our so-called civilization.

Spermageddon

Under the microscope
tadpole like human sperm
now swim in circles
instead of rushing upstream
to impregnate their target.

As viable sperm counts fall
and more sperm
swim erratically,
healthy swimmers may soon
become an endangered species.

To fight this impending spermageddon
men should be marching on Washington
to confront corporate lobbyists!
It's not us says the pesticide industry,
It's not us says the chemical industry,
It's not us says the plastics industry.

Collectively, these corporate hacks
point to the real villain,
the dreaded vegans
stuffing themselves
with soy products
that disrupt the endocrine system.

Star Man

Throughout eternity SpaceX's Star Man
gets to stare into the asteroid belt
feeling calm as a mannequin

For eternity he can heed
the dash screen's warning
"Don't Panic"

For eternity his cherry red Tesla convertible
will fly through space listening to
David Bowie's "Space Oddity"

Except no one hears it
as sound doesn't travel through
space's empty vacuum

If aliens encounter Star Man
they may question why earthlings
are such strange life forms

Note: This poem was inspired by Elon Musk's launch of his red Tesla convertible into space as a test of his SpaceX Rocket.

Another Planet

There's something distinct
about being old enough
to remember a time before
chattering cell phones,
rapid fire television ads,
constant email alerts,
abundant plastic debt,
and ridiculous selfie poses.

And ancient enough to remember
a time when "Duck and Cover" drills
warded off nuclear attack,
"Amazon" referred to a jungle,
and rock and roll
shattered segregation barriers.

So much changed so fast
this past must seem
like another planet
to over-stimulated young brains
constantly saving their planet
by annihilating Play Station aliens.

Happy Birthday Nature

At the end of the party
the festive balloons
are cut free and drift
into the nearest tree

Tangled in the oak's limbs
the sparkling mylar balloons
wait for a gust of wind
to jerk them loose

Then they can continue
over land to the sea
where the tides
exhaust the orbs

And rip them
into tiny shards
of mylar and string
impervious to decay

Curious about this shiny new food
fish and sea birds nibble away
until their corpse
becomes a monument
to forever birthdays

Drowning in the Internet

Help!
As we drown in a tsunami
of bits and bytes
some are spending as much time
on the internet as they do sleeping.

How can anyone keep up with
2.5 million Facebook posts,
300,000 tweets, and
72 hours of YouTube videos
uploading every minute?

In the evolution of consciousness
we have now reached such overload
that only Google maintains
our fragments of memory.

Mother Earth Meditation

Let's light a candle
in grief for:
      glaciers calving into the sea,
      rainforests disappearing,
      endangered species vanishing
      and souls lost in a spiritual desert.

Let the yellow flame
soothe our souls
and release our anguish
into the aether,
a place somewhere
between heaven and space.

As you contemplate the flame
imagine your contribution
to healing the planet.

Rebels of the Universe

Our universe is cold and dark
with occasional punctuations
of explosive celestial light.

In all this vastness
we may be rebel outliers
just odd organic creatures
gifted with the spirit of love,
compassion and hope.

As the odd bipedal creature
we are left to ponder
our fate and wonder
if we are alone.

Each Day

Crisis, extinction, shortage, catastrophe:
Words so overwhelming
we throw up our hands and
shout, "What's the use?

I'm only one of 7.6 billion
consuming the earth's resources,
polluting the environment,
carbonizing the atmosphere."

Building a habitable future
requires that each day
we undertake
one rebellious act:

Refuse a plastic item,
Plant a tree,
Walk or bike,
Conserve water,
Turn off a light,
Lower the thermostat,
Support peace,
or create your own dissent.

One little act each day times
7.6 billion hungry mouths
could make a big change.

[. . .]

You can become your own ad agency
and show what can be done
to sow seeds of optimism.

Then one day
our children
will thank us.

Culinary Consciousness

Viva La Tomato

With hints of seasonal warmth
the tiny seeds break open
in small peat pots
signaling an emerging spring

These seedlings can be lovingly placed
in the garden soil nourished
with compost, crab shells,
and rock phosphate

As calendar pages turn
yellow blooms explode on green stems
until like magic the flowers transform
into miniature green tomatoes

With spring giving way to summer's heat
the fruitlings grow in diameter
until hints of red and yellow
announce a culinary reward

When the tomato is fully ripened
the sharp knife's first slice
creates an olfactory explosion
that titillates taste buds

Relish the deep red slices standing
as a revolutionary counterpoint
to factory farmed imposters
oozing with impotent taste

Karmic Revenge

Native spirits must be smiling
as their ghosts peruse
empty grocery shelves
stripped bare by urban looters.
The barren shelves sit waiting
for just in time deliveries to cities
where no one can eat barcodes.

The massacred Eastern Shore natives
knew nature's abundance
and puzzled at immigrant whites
neglecting nature's rich larder of
plants growing by the river which could
make a meal of wood sorrel, dandelions,
wild onions, purslane and cattail stalks.

Native woven river weirs
snagged an abundant catch
of Striped Bass, Shad,
Mullet and Speckled Trout.
All supplemented with crabs
whose claws clamp tightly
to bits of animal flesh.

Hovering along the river Neuse
the massacred spirits of
Eastern Carolina tribes:
Tuscarora, Croatan and Catawba,
must enjoy the chaos as white tribes
fight over food shortages.

This lost knowledge
could be true karmic revenge.

Blueberry Mourning

It's a January frost-free miracle
in my North Carolina supermarket
where it's a two-for-one blueberry sale,
so I add them to my shopping cart.

Secure from the freaky outside cold
the Chilean bluets
look perfectly cozy
in their crystal-clear plastic containers.
They look so comfy, I wonder if they
still dream of summer in South America.

As I eat my morning oatmeal,
I ponder the adventure stories
of this well-travelled fruit.
Could it tell me about the toxic sprays
that made it picture-perfect?
Was it picked by a shaker machine
or by campesinos breaking their backs?

How was its 4,136-mile plane trip
from the Chilean farm to Florida?
Did it enjoy its 869-mile
truck trip up the interstate?
How many miles per gallon
does a Chilean blueberry get?

Sometime this summer
local pick-your-own blueberries
will ripen on nearby vines
and they can take a short cut
across the river to my morning bowl.

[. . .]

Then breakfast will be more enjoyable,
when my mind doesn't have to digest
so many perplexing questions.

The Perfect Holiday Meal

Planning the meal required great care
until relatives created a cook's nightmare
RSVP's came back with emails to heed
every individual's unique dietary need

The vegetarians asked for no meat
so that spared a turkey the oven heat
We planned to toast with eggnog
Until the lactose intolerant
wanted no seasonal dairy grog

So next we considered fresh fish
hoping it would make a pleasing main dish
A distant uncle worried about his ticker
became a real nuisance, that ol' stickler
I could hear the cook's bloodcurdling shriek
at the request of nothing with fins, face or beak

Then the exasperated cook let out a sigh
Saying at least there's fresh pumpkin pie
Until the next phone call added a request
Please nothing with sugar at my behest

That left only the homemade rolls
on a list that once resembled a scroll
Then a great aunt said prepare for me
a meal without wheat, a meal gluten free

Finally, one day we gathered at the table
For the meal that's now a family fable
The chef did his best to please every critic
serving a single fruit salad, free of anything acidic

Cooking Green

Almost like the prince
awakening Sleeping Beauty
with a fairy tale kiss,
collards are becoming gentrified
like kale its culinary cousin

Chefs now understand
the dark green plant strengthens bones,
prevents cancer, controls diabetes,
lifts depression, bolsters sleep,
and improves digestion

Hardy collards also survive
mild winters after frost
kills its vegetable friends
and it stands next to kale
as the sustainable queen

So, season up the pot
boil the collards down
sprinkle on some vinegar
and serve up a portion of health

Winter Smorgasbord

The California vegetables' long trip
finally comes to a grinding halt
in my juicer, separating pulp
to yield a nitrogen-rich mush,
for stoking my compost bin.

Under the snow-covered pile
Gala apple cores,
organic carrot peelings,
Minneola orange rinds,
and green cucumber skins
generate heat as they decompose
into an all-you-can-eat smorgasbord
for ravenous bugs, bacteria, and earthworms.

A generous dose of manure,
the Viagra of the soil world,
stimulates the little critters,
accelerating the breakdown
of leftovers and carbon-rich leaves
to transform the stack into a dark compost.

Like reincarnation, the compost
made from the well-travelled vegetables
becomes part of my summer bounty
of lettuce, peppers, tomatoes and herbs.

This home-grown produce will leave
no outrageous carbon footprint
in its short trip from garden to table.

The Downstream Loop

Sparkling in the sunlight
the little plastic bag sails
from the mindless driver's hand
to drift among roadside weeds

No one bothers to retrieve it
and on county mowing day
whirring blades cut a grass swath
shredding the bag into gossamer slivers

The next thunderstruck downpour
carries the shreds through the watershed
to the larger boiling stream
to the tidal marsh
to the Atlantic Ocean

The Gulf Stream's sun and waves
pulverize the slivers into tiny bits
creating a culinary delicacy
for large schools of filter feeders
small fish that mistake plastic globules
for aquatic eggs and plankton

Larger fish like Sea Trout and Tuna
cut a swath through the schools
and devour the tiny fish, concentrating
the petrochemicals up the food chain

For dinner we purchase the wild-caught Tuna,
let the fish monger filet the toxin-laden flesh,
pack it in ice, and store it in a virgin plastic bag
A bag that completes the ecological cycle

Sayonara Humans

To make a dab of honey
I have to sample nectar
from one hundred flowers.
And each pound requires visiting
millions of plants.

Now as our colonies collapse
from pesticides, parasites,
and climate change,
you will miss us
when we're gone.

We pollinate one third of your food
so, if you like your fruits, vegetables,
almonds, apples, strawberries,
blueberries, potatoes, peppers, and pears
get ready to kiss them goodbye.

Sayonara humans
eating just corn and soybeans
is going to get really boring.

Nature's Canaries

The Last Laugh

Imagine the sex life of bacteria.
No cumbersome courting for a mate,
just every ten minutes they divide
their single cell into two,
then double into four.

As masters of rapid adaptation
they flourish in hellish places:
acidic hot springs, barren soil, brackish water,
radioactive waste and the human gut.

In our body we have ten times
more bacteria than cells.
Probiotics are good for you
and germs are the bad ones.

Plain soap could do the job
but we use antibacterials
to cleanse counters, sterilize food
and sanitize the baby's butt.

They only kill 99 percent,
leaving you to wonder
if the one percent
plan a counterattack
with super resistant germs.

If bacteria could do a belly laugh
they'd be doubling over
because the real joke is on us:
humans who think we rule the Earth.

Deaf

As land animals we don't appreciate
how ocean creatures use sound
to find food, locate mates
and navigate the deep.

Imagine bombs going off in your
home every ten seconds
and you can understand
the sensation of air cannons
echo mapping for oil and gas.

Louder than a jet engine
the thousands of ocean blasts
create a tsunami of sound
confusing the echolocation
of whales and dolphins.

This torturing of ocean life
fails Genesis' admonition
to "replenish the earth,
... and have dominion
over the fish of the sea,
and the fowl of the air."

As this blasting of sound
threatens the marine web
we have to realize
we are the deaf ones.

Note: Originally written for Hands Across the Sea in 2018. At present, ocean blasting permits for seismic testing have expired and are now on hold as a result of delays created by citizen activism.

Elegant Travelers

In winter, I watch thousands
of lily-white Tundra Swans
dot Pungo Lake.

After their 4,000-mile flight,
the Alaskan travelers enjoy
North Carolina's milder clime.

From a distance, they look like
scattered snow clusters
on the water's surface.

In time a warming spring
will signal the elegant swans'
return flight to Alaska.

They never studied aerodynamics
but they stretch their necks,
fold back their lanky black legs,

form a V formation
and let flapping wings
reduce air resistance.

Flying at fifty miles per hour,
the flock of avian marvels travel
hundreds of miles each day,

before returning to Alaska,
where life partners hatch offspring
and prepare chicks to migrate South.

[. . .]

But it saddens me to know
Pungo Lake will one day be barren,
as global warming moves
their winter destination north.

Translucent Canaries

Around the rocky trail's bend
flashes of orange flicker
like Smokey Mountain fire,
ignited by thousands of migrating
Monarch butterflies recharging
for their flight to Mexico.

So light and delicate
how could these orgasms of color
navigate thousands of miles
to their ancestral tree?

Monarchs only lay their eggs
on scarce native milkweed.
As these eggs hatch into caterpillars
they gorge on the plant's leaves.

Then hang upside down,
and perform a rhythmic dance
of squeezes and pulsations
to mold a chrysalis cocoon
that transforms the "cats"
into a royal adult.

Repeated hatches over the summer
give birth to multi-patterned Monarchs
but only the last hatch flies
thousands of miles south
to a winter home.

{ . . . }

Today, like a virtual canary
in Nature's coal mine,
fewer Monarchs make it
to the Mexican forest to await
a spring warmth that will awaken the hibernators
for their flight north on translucent wings.

Nature's Bacchanal

On a summer evening
we are assaulted with a buzz
so loud it echoes in our ears.

The maddening sound heralds
the return of cicada broods
from their subterranean lair,
where they've spent years
sipping tree root nectar.

Some broods wait over a decade
underground dreaming
X-rated cicada dreams
until their internal clock
announces its time to break free.

Normally the broods are predictable
but climate change confuses
their biological clock
so this year they emerge early
for a ten-week raucous mating party
which gives new meaning to a quickie.

Lightning Bug Apology

Dear Lightning Bugs,
On behalf of all those who
are not aware you face extinction
I would like to apologize.

As you disappear along
with the bees, amphibians,
and Monarch butterflies
I apologize for

our attempts
to reorder nature through
habitat loss, toxic chemicals
and light pollution.

I apologize for the night lighting
of streetlights and headlights
that disrupt the few weeks
your light blinks to attract a mate.

What a pity, many will
never know the joy of watching
your fireworks
outshine the stars.

Leaf Meditation

Leaves are the lungs
of Mother Earth.

Now close your eyes
and visualize a forest.
Find just one tree
and bring your focus
to one leaf on the tree.

As you exhale CO_2
this leaf breathes it in
to create the oxygen
we inhale.

As your mind focuses
on the reciprocal breaths
contemplate your role
in the web of life.

Hummingbird Laughter

The oily seeds in my backyard feeder
create a regular dinner party
for black feathered Grackles.
These sloppy, aggressive eaters
knock morsels to the ground
so a lower caste of brown suited Doves
can scavenge downstairs for leftovers.

With the dinner party in full blast
a fire red Cardinal flies in
like a Kung Fu fighter
to scatter the large black bullies
and gorge himself on the feeder's seeds.
His less colorful mate patiently waits
her turn to savor sunflower seeds.

After they dine, the Kung Fu Cardinal
clears an opening for more delicate guests
like common Sparrows, Towhees,
and the belle of the ball, the House Finch
sporting bright colorful plumage.
Once a living room star
it escaped its wire framed cell
to mate with others in the wild.

Nearby at the liquid feeder
a Hummingbird hovers,
laughing at the seed eaters
whose clumsy beaks can't savor
the bliss of sweet flower nectars.
Between sips it chuckles
at evolution that freed it
in the competition for top bird.

Mourning Dove Spring

My closest neighbor
is a Mourning Dove
secluded in the evergreen holly
on a straw and twig nest
outside my bedroom.

I felt sorry for the mother
sitting alone on the nest for weeks
until I learned the gender doppelgangers
take shifts incubating two small eggs.

In about two weeks
little tufted grey heads pop out
to innocently look around at their new world
oblivious to danger from hawks,
snakes, squirrels, cats and
their most aggressive predator—hunters.

Just two weeks after the hatch
the young are nudged out of the nest
to drift down and test their wings
with short aerial hops
bouncing on the ground.

With the nest empty
Mr. Mourning Dove
puffs out his breast, and
initiates a mating dance
to start the cycle over.

Back in their barren nest
the two lovebirds
restart the spring ritual.

Lunacy

The Romans worshipped Luna
as a goddess who births a new moon
guards the feminine realm
and renews monthly fertility cycles.

In the guise of a slivered moon
she is the maiden goddess,
fresh and virginal
symbolizing a new beginning.

Then as her fullness increases
she becomes the mother goddess
pregnant with an abundant life
and infinite possibilities.

In time her fullness wanes
transforming her into a crone,
an ancient guardian of magical arts
and a source of wiccan power.

At the dimming of nocturnal light
humans fear the lunacy will release
werewolves and wombats
to roam the eerie night.

Soon a new moon
pierces the darkness
and heralds a new cycle
to celebrate a new beginning.

Ghost Stumps

My laminate counters, dirty and cracking,
are no longer the pinnacle of kitchen style.
They are alien in my historic home,
a real Southern Victorian lady,
crying out for restoration.

The house's heart pine bones stand strong,
a tribute to over 100,000 square miles
of ancient trees that once rose arrow straight
into the Southern sky.

Centuries of slow growth
produced a steel-hard wood
impervious to mold, insects and rot,
but not to the boiling water
powering razor-sharp blades.

Natives respected the forest that
sustained them, but where natives
saw sustenance white men
imagined product yielding profit
from iron-hard lumber.

Lumber to build:
wharves,
factories,
homes.

In mere decades, the ancient woods
vanished, leaving only ghost stumps
watched over by faster growing
soft pines.

[. . .]

Today old mills constructed
of heart pine, sit silent until the day
salvagers strip the hollow factories
of "antique" boards, hard enough
to create a kitchen counter
for my Victorian mistress.

Long Leaf Cathedral

Consider the boards in my old house
pine boards much wider than
the little sticks in houses today.
Looking at the aged lumber
it's hard to imagine the parent,
a mammoth Longleaf Pine.

Before Europeans arrived
Longleaf Pine forests
blanketed 90 million acres
creating a natural cathedral
of 500-year-old trees
soaring more than 150 feet
above the Southern forest floor.

These towering trees
of iron hard lumber
impervious to rot, mildew and decay,
birthed a timber gold rush
that leveled the forests.

Finding a mature specimen
today is a rare gift
but on occasion
a ghost of the ancient forests
soars before me—a giant
that somehow survived the slaughter
of the lumberman's shark toothed saw.

Chasing Venus

Somewhere deep in a scruffy
long leaf Carolina coastal bog
a threatened monster plant thrives
guarded by poisonous snakes,
tiny ticks and biting chiggers.

Envisioning the image
of a Little Shop of Horrors
giant meat-eating plant,
I was surprised to see a small plant
with a one-inch red trap
open to lure insects inside.

On the insect's first touch
of trigger hairs the trap
is primed and on the second brush
the clam shell snaps shut
trapping the unsuspecting insect
behind hair like blades.

Unlike most plants Venus Fly Traps
are carnivorous meat eaters
that secrete digestive juices
to dissolve an insect's soft parts
before ejecting the skeletal remains.

Early settlers thought this
odd plant's royal crimson trap
mimicked a female's genitalia
so they named it Venus
after the Roman goddess of love.

But the wild Venus Flytrap
isn't very lovable to the insects
lured inside with fragrant nectar.

And unlike the Venus Fly Trap,
a woman's private parts
never evolved to snap shut
and lock their intruder
in hair like appendages
making escape impossible.

Riding It Out

2011

Like a horror movie trailer
we watch the computer screen
for signs the emerging hurricane
will shift before it barrels straight for us.

Soon the screen's colorful swirls
make our trees bow to the wind gods.
With the storm creeping closer
we watch on the electronic altar
and whisper the name Irene.
We feel a bond with the Ancient Greeks
who begged mercy from Eurus,
the eastern wind god.

On the internet radar, we watch
the storm's tentacles spin counterclockwise
picking up salt water that kills our garden.
As torrents of rain pelt our house
Tung nuts blast the sides like random bullets.
Then an errant projectile breaks a window
scattering shards throughout the room.

The intense wind splinters tree limbs
that sever critical power lines.
Driving rain penetrates the roof and walls,
leaving darkened rooms covered in blue tarps
adorned with pots and pans
to catch the dripping water.

Just as it seems unbearable, the gray sky
lightens and calm settles over the house.
But then the storm's eye moves north.
and a driving rain peels back
a neighbor's roof to create
a waterfall over his grand piano.

As the floodwater rises to the edge of our porch
a neighbor's foundation becomes a depth gauge.
Not bothered by the water, our dog swims up the block
in search of a grassy island to relieve himself.
Doggy paddling back, he smiles,
thinking this new river is awesome.

In time the flooding subsides
and the river returns to its banks
leaving stranded fish to flop in puddles
until a neighbor scoops them in a bucket
and returns them to their aqueous home.

After the nightmare storm ends,
neighbors fire up smoky grills,
communally cook freezer-thawed food,
and raise a survival toast
to the wind gods journeying north.

Mother Nature Is a Bitch

2018

With warming seas
an emerging tropical storm
evolves into a major hurricane
and births Hurricane Florence.

Like a nocturnal monster Florence's
thirteen-foot storm surge
slams onto our porch bringing water
to the bottom of our windows.

The front door oozes water
and the first floor turns
into a three-foot-deep
indoor pool.

The torrential wind and rain
start a few small roof leaks
which over two days open up
like a festering wound.

After days of rain
electricity and air conditioning
become a technological fantasy
as the closed house feels
like being locked in a sauna.

With 100 percent humidity and
drenched boards the house swells
and doors freeze shut so tight
you wish looters would pry them open.

Once pried open
the doors don't close
and locking the house
becomes a security novelty.

With so much moisture
the slugs move up the windows
and water bugs
retreat to the ceiling.

With tap water still running
I endure a cold shower
that slams my body parts
like a gust of arctic air.

Music books downstairs
that promised to make me
a banjo virtuoso drown
like my lack of musical talent.

Why did I ride out the hurricane
in a century old house?
Maybe I like the chaos of Mother Nature
showing us who is really in charge.

A hurricane like Florence
reminds us it's Her planet
and cautions us not to build
new houses on wetlands.

But as the streets drain
the cleanup starts
and we go back
to showing the bitch
who is boss.

Recovering Hope

Dark lines now mark
Hurricane Florence's tidal surge—
a wave that undercut foundations,
destroyed doors, flooded houses.

Water-soaked walls now culture black mold
behind bead board and sheet rock
that can only be attacked by ripping open
the wall with an army of pry bars.

Mountains of refuse rise
street side, like some monument
to the building materials
that once shielded us from the elements.

The piles remind us water
is a hydra-headed monster
that can both nurture and destroy
with enough power to break boats loose
and float them onto manicured lawns.

Surveying the shattered boards,
mangled ductwork,
soaked furniture and ripped insulation
rebuilding seems so overwhelming,
hope becomes a casualty
of the storm-tossed nightmare.

But then, on the monochrome pallet
of our shattered dreams,
a Monarch gently lands on a splinter
of century-old pine and sips a dewdrop.

The butterfly's bright wings offer a glimpse
that tomorrow the sunrise will color the landscape
and herald that new hope can rise from the rubble.

Unleashing the Cosmic Jokers

As Hurricane Florence's waves
ripped down our street
Mother Nature didn't bother to knock
and doors, windows, foundations
buckle under her wraith.

Surveying the damage
from the winds and tidal surge
some throw up their hands
turn their backs
and walk away.

For those who stayed
the hurricane released
spiritual trickster gods
like the Native American's Coyote
who relishes pranking us.

Coyote must now be laughing
as overloaded contractors
stop returning phone calls,
electric companies pull meters,
corroded waterlines break, and
building materials sell out.

In considerable agony
I discover life can exist
without the internet, gas heat,
and electric power.

Staring at a crippled kitchen
I start to fantasize
a working sink with hot water
a microwave as the new hearth
and two sawhorses as my table.

With deep breaths and slow exhales
I realize I am lucky
because in spite of the losses
I have a house to rebuild
unlike those who lost their homes
in a Florida hurricane
or a California forest fire.

So, as I contemplate the restoration
I will flip the bird
at the trickster Coyote
and grin at the universe.

Deconstructing the Future

Awakening

No one hears the solar rays
as greenhouse gases trap heat
to toast the planet

Until the tears of Mother Earth
pelt down as torrential rains

Until glaciers crack under
the Arctic's eternal summer

Until sea-warmed hurricanes
undercut oceanside foundations

Until verdant forests crackle
with drought-nourished fires

Until habitats disappear and
species cry out for a new home

May this dystopian symphony
awaken us from our suicidal slumber

A Tiny Acorn

The forest's monster oak
towering over the canopy
starts as a tiny acorn
burrowed into the soil.

Greta Thunberg began alone
outside the Swedish Parliament
with her hand lettered sign
crying out, "Our house is on fire."

Fearlessly she viewed her
OCD as a gift to challenge
world leaders with
 "How dare you
for stealing our future."

Like the oak's tiny seed
her message sprouted
and rippled around the planet
in a movement politicians
can no longer ignore.

Let's hope her urgency
spreads to other issues
and peace breaks out
in a de-nuclear world
where we smile at the sun.

Forest Bathing

My residential cocoon
is really cozy as it
guards me from
nature's wildness.

My illuminated habitat
wards off the elements
and creates its own micro climate
oblivious to its carbon footprint.

And yet something is missing
as artificial light challenges
the setting sun and stale air
maintains a constant temperature.

In contrast a short distance away
nature beckons me to a forest
where natural bioenergy
can alter my mental state.

Strolling through this verdant space
I enjoy a heightened awareness
of life's web and become open
to unspoiled wildness.

Feeling restored I thank the trees
and say goodbye to the
rustling leaves, trickling water,
melodic birds, dappling light,
and healing spirits.

Note: The Japanese believe time in the forest can be healthy. They practice forest bathing, or *shinrin-yoku*.

Homo Sapiens' Abridged History

Programmed in our DNA
is a fetish to cleave
the world around us.

First rocks crudely smashed
stone, flesh and soil
until flint's edge yielded
sharper, easier cuts.

Then, a Neolithic pottery kiln
birthed the bronze age
by mixing rivulets
of copper and tin alloys
to forge destructive implements.

Implements that could split heads,
fell trees and forge plows
to rip the earth.

Early in the Twentieth Century
we started cracking oil, and natural gas,
to create gasoline and plastics.

This led to explosive demand
and humans ravaged land and sea
for more hydrocarbons
that now smother the atmosphere.

Highway Dreams

Parallel to the manicured four lane
a narrow two-lane road
snakes past gravel entrances
to a broken dream.

Rusted gas pumps sit idle
with mechanical numbers frozen
in a time when a gallon of regular
sold for less than a dollar.

Rickety wooden stands
for fresh-picked fruits and vegetables
sit empty, abandoned by farmers
undercut by Wal-Mart's foreign produce.

Faded hand-painted signs
announce antiques and mountain curios
once lovingly crafted by locals
who never called themselves artisans.

Neon filled electric signs
will never again light up
to announce the café is open
to serve hot coffee and fresh-baked pies.

These buildings with peeling paint,
rotting wood, and broken windows
stand as historical monuments
to a local person's broken dream.

[. . .]

An independent country dream
of owning your own business
before the interstate stole customers
now flying by at seventy miles an hour.

A Monument to Another Time

A winding rutted road
rambles through scattered rocks
to an abandoned homestead
that traces time backwards.

In the overgrown clearing
a hand laid stone chimney
pokes above winding vines
and gnarled tree limbs.

The fireplace stands as
tribute to an unknown mason
whose calloused hands
meticulously stacked the stones.

With the charred house gone
front porch music
no longer blesses the mountain
with notes and harmonies
that surf the Appalachian wind.

In spring wild flowers
scatter sun dappled beauty
among the crannies of this dream
of a simpler life, an abundant garden
and a small homestead taming nature.

Through winter the chimney
stands alone among
a pallet of brown hues
that wait for spring shoots
to burst forth and repaint
the landscape.

Crossing the Void

On the Starship Enterprise
space is the final frontier
but closer to planet Earth
the racial voids that divide us
present the real challenge.

Peace and harmony require
crossing this chasm
to see the world through
the blinders on others' eyes.

To look beyond the slogans
armored with glass ceilings,
racial bias, sexual harassment,
inequality and bloodied histories.

With the planet's seven billion souls
longing for meaning and sustenance,
the chasms between us are more dangerous
than the black holes of space.

In this overcrowded world
survival requires crossing the void
so pack your emotional baggage and
blast off across the unknown.

Visiting Khatyn

Peace Memorial in Minsk Region, Belarus

At sunset each step up the earthen berm
slowly reveals stone chimneys standing
as monuments to an unimagined darkness
that reduced hundreds of villages
to stone rubble and ashen timbers.

Across the field masonry memorializes
thousands of villagers burned alive
as Fascists sought revenge
for partisan guerilla attacks
launched from surrounding forests.

On hearths reaching to the horizon
urns rest filled with ashes and soil
scooped from the 628 flamed hamlets.
Each now lovingly stands as
a spiritual reminder of war's insanity.

Three solitary birch trees and an eternal flame
symbolize the one quarter of Belarusians killed
in the world war that targeted their villages.
On this site twenty-six bells toll every hour
to remember the homes that once stood here.

The wind that whipped the flames
and charred the flesh, now cleanses the earth
leaving only spirits to haunt the memorial
and remind us of the horrors of war.

Redefining Habitat

Without shells and fir
our species compensates
by constructing rigid walls
from earth, wood, stone and plastic.

In our oversized armored habitats
we ward off the elements
by creating a micro climate
oblivious to its surrounding impact.

We condition the air
by burning fossil fuels
and exhausting the gases
to add to global warming.

Wouldn't it be wonderful
if our dwellings could be more
like dynamic organisms
with a surface responding
to seasonal flux?

Perhaps a plantlike surface
could change state to reflect
solar heat in summer
and absorb it in winter.

This smart stomata skin
could open and close
to regulate internal humidity
and transpire it to the outside.

Such natural technology
could make our habitat
one with the trees and plants
and redefine home.

Ballet Mechanistic

Like a choreographed dance
the gunmetal arms swing,
rotate, lift and swoosh
following targets of laser eyes
plotting points X, Y and Z

Spraying, welding, and machining
with precise moves once done by humans
the robots' limbs swing back and forth
and are never bored, never demand a break
and never walk a picket line

Complete with opposable claws,
flexible elbows and pressure sensors
the bots position components
with a delicate touch
more consistent than
a human's grip

Mimicking repetitive motion,
workers now train robots
to take their jobs by
matching their moves

As the robots go through their fluid ballet
one has to ponder if the chipped brains
are simply commanding movements
or if they dream of a self-replicating world
freed from human commands

Big Blue Monster

Early one winter morning
the future came rumbling
down the side of my street
in the guise of a garbage truck

Pausing by the curb
the truck's only occupant
extends a giant mechanical claw
to the waiting tall trash receptacle

Its huge fingers straighten
and fondle the plastic container
with vice-grip delicacy
before flipping it into the air

At its zenith the lid flops open
and dumps a week's assemblage
of plastic wrappers, tissues,
food waste, and assorted trash

Once upon a time one driver steered
and two workers hung off the back
to wrestle the large refuse bins
with the agility of circus performers

A nimble human finger would press a button
to compress the trash into a sculpture
of found art displayed briefly before burial
at the county's overflowing landfill

[. . .]

So next time you hear politicians
talk job creation, think about the ghosts
in bright orange jump suits
who once hung off the truck's back
until the robot claw underbid them
and put their families on food stamps

How Many Falls?

South Toe, North Carolina campground

The burst of colors, chilly air
and rushing water's white noise
open a door to the hidden spirit
of past autumns along the South Toe.

Camped by the flowing water
you question if any of its large trees
stood witness to Indians landing
their canoes at their winter camp.

Did the oak spy on Confederate deserters
hiding out from home guard bounty hunters
or watch the subsistence settlers feverishly
can vegetables, smoke trout and skin deer
to survive winter's isolation?

Is the ancient cedar feeling lucky
to have escaped the burly loggers
who cleared the mountain forests
and drifted their fallen prizes down
the swollen summer river?

Did the maple feel guilty
when the surveyor saved it
by marking its neighbors
for sacrifice to create
large concrete RV pads?

[. . .]

RVs so homey, occupants can watch
the National Geographic channel
and observe two-dimensional nature
on their forty-eight-inch satellite TV.

Is fall now more festive for the cedar
as campers decorate their campsite
with little glowing Chinese lanterns
that dance in the wind,
and cast a colorful orange glow
on Astroturf front lawns?

The Rebellion

In a suburb far far away
it is now the latest fashion
to dress pet dogs
in cutesy holiday costumes

And post Facebook pictures
so other pet lovers can ooh and aww
at dogs dressed with dinosaur tails,
lobster claws, bow ties, Christmas hats,
antlers, superman capes,
and bumblebee wings

But now even the dogs costumed
as Darth Vader and Princess Leia
sense a disturbance in the force
and to fight the pet store empire
Scarface, the pit bull rebels

In a legendary moment he
takes a stand for all
canine holiday foolishness
by resisting his Christmas sweater
and bites his owner to send
other households a message
that dogs prefer *au naturel*

Grass Meets Its Match

What better testosterone high
than to make a lawnmower really fly
down the straightaway at a record speed
pushing the driver to take the lead.

Sleek, mean and fire engine red
it's the machine that grass will dread.
as it roars down the track
powered by a 1,000 cc motorcycle Twin.

With a 116 mph documented speed
Guinness made the company really exceed
as the world's fastest mower
beating others much slower.

Yet, in mowing grass it will still be slow
only 15 mph with spinning blades below.
On the track is where it can really shine
beating mowers of more impotent design.

Note: In 2014 Honda's racing team modified a Honda lawnmower to challenge the Guinness World Record for fastest riding lawn mower. It beat the old record by 30 miles per hour.

Apologies to a Tree

My fingers fly across the keyboard
converting bursts of mental images
to words destined to drown
in the digital tsunami
flooding the internet.

Internet moguls offer to store
my ramblings in a server farm,
but my gut tells me
clay tablets and paper
will outlast the modern cloud.

Now technology is so vulnerable
one employee can program delete,
code a worm in the back door,
or overload servers
with misspelled Viagra offers.

So, with whispered apologies to trees,
I hover the mouse's cursor
over the icon and click print,
betting printed words will survive
a solar electromagnetic pulse
that can fry computer storage
and wipe clean the digital age.

Grey Dawn

Dense fog paints the river trail
with gradations of black and white
leaving me to walk among
a limited grey pallet.

With fog diffusing the morning light
and dampening the shadows
I walk through a visual illusion
void of depth and dimension.

With each uncertain step
I walk knowing the sun
will soon burn away the fog
and clarify my future path.

Evolution's U-Turn

News headlines crawl
like an emotional jigsaw puzzle
across the bottom of the screen
as a testament to a world coming unglued:

- North Korea threatens a nuclear attack
- Summer heat wave sets new record
- Politicians propose impeachment
- Cats are still on menus in Vietnam
- A new Covid variant emerges
- Amish wives develop a taste for yoga
- New York's streets showcase nude art
    and
- Khloe Kardashian takes a Ramadan sex break

So much for evolutionary progress.

Monkey Mind

I can still see the Buddhist artist
in an orange robe quietly
sitting on the national mall
painting under a large oak tree.

His muted brush strokes
bring life to a line of elephants
ascending a winding mountain trail
with a single monkey riding precariously
on the back of each pachyderm.

As each elephant climbs the mountain
its monkey becomes less visible
until only a faint outline
is perched on its broad back.

A man in madras shorts asks
"Does this painting have a meaning?"
Patiently the open-air docent replies,
"The elephants' journey represents
our mind filled with chattering monkeys
and with each meditation breath
the monkeys become quieter."

Restoring Consciousness

Liberate your mind
cut the cable
de-dish the dish
celebrate the off switch
restore pre-TV consciousness

Imagine a world where
producers don't addict us
twenty-four hours a day
to blood, gore and sex

A world where children
aren't bombarded with
two hundred and fifty thousand
VIOLENT images before age 18

A world of natural stimulation
of chirping birds,
fire orange sunsets,
and star speckled night skies

Buddhists imagine the mind
a cage of chattering monkeys
where meditation can open
the creaky cage door
to let the monkeys escape

Then the mind can become
so peacefully quiet
the resonance of past lives
will whisper in our ears

Sitting Still

Travelling thousands of miles
leads to a small village in India
where the guru instructs us
to learn to do nothing

Sitting still is harder than it looks
for an overloaded Western mind
unable to concentrate
and continuously dancing
through time and space

Reshaping the Future

Social change starts
when the abuses
ignite a smoldering fire
in our collective soul.

History written by victors
only creates a veneer
of collective truth
that inevitably cracks,

And the masses begin
to demand change in:
       global warming,
       simmering racism,
       rampant sexism,
       concentrated wealth.

Social change is like a dormant volcano
with subterranean pressure building up
until the lava suddenly erupts
and reshapes the mountain.

Story Power

Long before stories
flew through the universe
as fractured digital bits
stories moved people
to love,
to war,
to help.

Researchers show stories
igniting brain receptors
to create empathy,
to stir emotions,
to trigger change.

Good stories stick
to cultures like glue.
so it is no wonder
Jesus told stories.
Buddha told stories.
Muhammad told stories.

Facts just get lost in the clutter
but with well-crafted survival stories
we could imagine a viable future.

Earth's Epilogue

Rust

No matter how smart
we think we are
entropy always rears its head
like nature's joker
reminding us we are never
long term smart.

We may have gained an edge
over other primates
with our molten metal forged
in fires hot enough to shape iron
into weapons, tools, and trusses.

Industrial age artifacts
now litter the landscape
as refuse sculptures
covered in a brown
patina which will soon
pulverize into metallic dust.

In our latest attempt to outsmart nature
we are substituting plastic for iron and steel,
a new material that never corrodes
but only breaks into smaller bits
readily absorbed by natural systems.

Unlike rust which is not digestible
our chemistry worship
could soon come back
to bite us in our biological ass.

Rusted Dreams

~on visiting an antique car junkyard

The material world offers
a plethora of automotive products—
V8's to enhance our power,
leather seats to soothe our soul,
and convertibles to lift our spirits.

These sparkling consumer objects
become shimmering mirrors
that reflect the false happiness
of our fatigued souls.

In time the mirror becomes cloudy
and rust corrodes our shiny dreams
revealing nature's antidote
to prosperity's false promise.

One Word: Plastic

As wood, metal, bone, and tusk,
became scarce, humans couldn't resist
a utopian future material.

A chemical synthetic wonder
that could realize a designer's wildest shapes
for steering wheels, hair ornaments,
and colorful radio cases.

This revolutionary material for a new age
could mold inexpensive petrochemicals
into bold pearlized colors: Rose,
Teal Green, Sunrise Yellow, Lilac.

The industry heralded
better living through chemistry
and sold plastics as an antidote
to our ravaging of the natural world.

Seduced by this modern marvel,
throwaway plastic exploded
littering Earth with bits now
finding their way into our bloodstream.

If aliens wanted to destroy rival humanoids,
what better way than to dangle
a synthetic material so enticing,
we couldn't resist the lure
of a plastic covered Earth?

City in Grey

The shining city's concrete
and glass skyscrapers
reflect a decaying grey patina cleansed
only by occasional white snow
falling on thick glass walls
shielding occupants from nature

On the edge of the city
the terrain burns a yellow glow
as machines gnaw the earth
to produce fuel to sustain
the soulless architecture
that slices the murky sky

In this megalomaniacal negation
of human-scale neighborhoods,
Mumford's walkable garden city
is bulldozed to create Le Corbusier's
sterile vision of straight concrete lines,
residential towers, and sixty story offices

Nearly Natural

~cemetery ad copy

What could be lovelier
than a field of colorful flowers
adding a touch of near-nature
to acres of graves?

Permanently dyed artificial flowers
will display your love for months
and complement the mortician's work
who pickled your dearly beloved.

Let your creativity go wild
with unearthly purples, pinks, and greens
adorned with a beautiful butterfly,
securely adhered to resist the elements.

For a little extra money
let an aerosol sealant
keep rain, wind, and sun
from fading your silk flowers.

Or you can distinguish a grave
with solar powered LED lilies
that recharge in the day and greet
the evening with a soft, heavenly glow.

Remember, nothing makes a grave
stand out from the stone-cold crowd
like nearly-natural synthetic flowers—
flowers that keep standing up
when your loved one lies down.

Forever Green

My wife always wanted mountain property
forests, a stream, clean air, wild flowers
a peaceful spot off the beaten path
a real forever home.

A place where you can get in touch with nature
and nothing can put you in touch with nature
for eternity like a green burial
at Carolina Memorial Sanctuary.

You can be peacefully
laid out in:
a biodegradable shroud
a recycled cardboard box
or an unfinished wooden box.

You can rest comfortably
knowing your burial
will contribute to annually saving:
77,000 hardwood trees,
1.6 million tons of concrete,
and 81,500 tons of metal.

At the Sanctuary your spirit
will be at ease knowing
you saved energy
and didn't pollute the air
by choosing cremation.

Your spirit can rest peacefully
as you provide a feast
for living organisms including:
bacteria,
worms,
microorganisms.

In the end
think of yourself
as a holy compost
so that all that will be left
is a small stone and
a GPS coordinate.

The Ecology Symbol

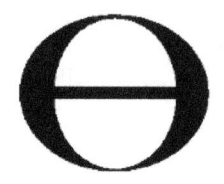

Once upon an Earth Day,
millions of marchers displayed
Ron Cobb's design,
melding the E for environment
and the O for organism
to create the ecology symbol.

Such a simple graphic,
just a circle and slash to symbolize
care for the planet,
respect for nature,
and the nurturing of a legacy
for generations unborn.

I don't see the ecology symbol
at today's climate marches
but it's co-conspirator, the peace symbol,
seems to be everywhere.
At Wal-Mart you can buy it
on underwear and t-shirts
that glow in the dark.

The vanishing ecology symbol
must be too threatening
to the dollar sign worshippers,
with its pesky admonition
to reduce consumption,
reuse materials, and respect nature.

It must be too threatening
to the comfort of North Americans
who consume 60 percent
of the Earth's resources
just to support our obese lifestyles.

It must be too threatening
to the 80 million new babies
each year who will aspire
to an American lifestyle.

Soon they will discover
if everyone lives the American dream
we will need a planet three times
the size of Mother Earth and
the last time I looked,
she's not getting larger.

Unleash the Dove

In many religions the dove
is a harbinger of renewal.
After the flood Noah
released a white dove.

It returned with an olive branch
to reveal the discovery
of a verdant land where
he could start anew.

In these tumultuous times
let's unleash the dove of peace
to search out alternatives to
exploitation, hunger and war.

Hopefully it will return
with an olive branch
to announce the good news
of the dawning of a new day.

A Perfect Legacy

Celebrate your new home
by planting a tree,
transplanting a shoot
or burying a tiny acorn.

In time, a strong limb
will be there when children
need to soar on a swing.

Let the trunk reach up,
so limbs can block the summer heat
and in fall the colored leaves
can nourish rich compost.

In the branches, birds
and squirrels will
weave their nests.

And one day you will be celebrated
by neighbors and family
as the one who sowed the seed
and nurtured the sprout.

Acknowledgments

Some of these poems have appeared in various publications and I have benefitted from editorial comments from editors including Isabelle Kenyon with Fly on the Wall Press, Judyth Hill, performance poet and writing instructor, and the editors at Unsolicited Press. I also need to thank my wife, Mary Anne, who gives me occasional feedback.

Gratitude and acknowledgment to the editors of the following journals where these poems were first published:

The Lyricist: "The Perfect Holiday Meal"

Duke University's *Eno Magazine*: "The Downstream Loop," "Blueberry Mourning," "Turtle Earth," "Viva La Tomato," "If We Could Hear the Earth," and "Tiny Acorn"

Purple Breakfast Review, London England: "Sitting Still"

Flying South: "Ballet Mechanistic"

Slippery Elm: "Apologies to a Tree"

Sleet: "The Ecology Symbol"

Poetry in Plain Sight (broadside): "Another Planet"

Kakalak (2019): "Unleashing the Cosmic Jokers"

The following poems also appear in my book, *Cogitation* (Unsolicited Press, Portland, Oregon):

"Restoring Consciousness" "Translucent Canaries," "The Last Laugh," "Sayonara Humans," "Your Legacy," "Rusted Dreams," "Ghost Stumps," "Rebels of the Universe," and "Highway Dreams"

The following poems appear in my chapbook, *Awakening: Musings on Planetary Survival* (Fly on the Wall Press, Manchester, England):

"The Web," "Awakening," "Jacuzzi Guilt," "Winter Smorgasbord," "Elegant Travelers," "Karmic Revenge," "Spermageddon," "One Word: Plastics," "Nearly Natural," and "Urban Tumbleweeds"

"The Downstream Loop" evolved into the award-winning, illustrated children's book, *My Little Plastic Bag* (2016)

About the Author

After working as an environmental advocate Sam Love made a living creating film and video images. With so much internet visual clutter he decided to transition to poetry where readers can make the movie in their heads. Sam now lives in New Bern, N.C.

Fly on the Wall Press in Manchester, England published his chapbook *Awakening: Musings on Planetary Survival*. His poetry collection exploring cracks in our culture, *Cogitation*, is available from Unsolicited Press and his self-published illustrated children's book *My Little Plastic Bag* has won numerous awards including a Nautilus Book Award. It is available in Spanish and English.

In 1970 Sam worked on the national staff of the first Earth Day. He was a founding editor of the successor group's magazine Environmental Action and served as Coordinator of the organization which worked to transform the momentum created by Earth Day into legislative changes. Over the years he published numerous articles in mass circulation magazines including *Washingtonian* and *Smithsonian*.

His poems have appeared in *Kakalak, Slippery Elm, Voices on the Wind, The Lyricist, Flying South, Sleet*, and other publications. *Eno* published by Duke University published six of his environmental poems and four were featured on Poetry in Plain Sight posters exhibited throughout North Carolina.

www.samlove.net
www.mylittleplasticbag.com
twitter @samlovepoet

About The Poetry Box

The Poetry Box® is a boutique publishing company in Portland, Oregon, which provides a platform for both established and emerging poets to share their words with the world through beautiful printed books and chapbooks.

Feel free to visit the online bookstore (thePoetryBox.com), where you'll find more titles including:

Exchanging Wisdom by Christopher & Angelo Luna

The Catalog of Small Contentments by Carolyn Martin

The Shape of Sky by Cathy Cain

Like the O in Hope by Jeanne Julian

Bee Dance by Cathy Cain

Stronger Than the Current by Mark Thalman

What Is Not a Miracle by Don Badgley

Between States of Matter by Sherry Rind

Sophia & Mister Walter Whitman by Penelope Scambly Schott

A Long, Wide Stretch of Calm by Melanie Green

World Gone Zoom by David Belmont

What We Bring Home by Susan Coultrap-McQuin

and more . . .